Axe-It-First

Preventing Catastrophic-Mega-Wildfires

By

Ron Rommel

AXE-IT-FIRST

Targets a Plan to Prevent Catastrophic-Mega-Wildfires

(Katu.com 2021) Author accessed March 2022

Year 2021 Bootleg Wildfire, Oregon's 3rd largest since 1900

©2022

ISBN 978-0-9862369-3-8

All Rights Reserved Including the right to reproduce

This Primer Report or parts thereof in any form.

All Images, Tables, Graphs, Diagrams, and Illustrations are either

Website Public Domain or have permission to copy within the Cited Works.

All Photographs are either Website Public Domain or have permission to copy within the Cited Works

Or

Photographs are either photographed or granted to the Author.

Dedication
Axe-It-First is dedicated to:

- The people of local wildland firefighters and to those who lost the battle.
- The folks who serve as Fire Lookouts and share their eyes to alert fire response teams.
- Dispatch and support personnel who serve multiple agencies to battle wildfires.
- All serving government, tribal, corporate, private, and the wildland-urban-interface.
- All who seek a balance to protect, preserve and adapt our forests to Climate Change.

Loving thanks to my wife, Liz, for proofing this critical Primer Report.

Thanks to Pamala J. Vincent, Editor and Self-Publishing Author for this finished book.

If it were not for Pam, this book would have never reached you, my audience.

(Barnas and Forest 1980)

Photo taken in 1980 at Olallie Lake Resort for meticulous Fire Prevention.

Author, Ron Rommel built "Smokey Bear Plaque" and presented to

Resort Operators by Ron on far left and Ladies from (Recreation) on right.

Olallie Lake Resort is in Oregon on the Clackamas Ranger District, of the Mt. Hood National Forest, U.S. Forest Service (USFS).

Acknowledging those who embrace Education and Implementation:

Oregon Department of Forestry (ODF), Forester Randy Baley for his interview.

Thanks, Randy for your participation in the Klamath Collaborative and your insight of the 2021 Bootleg fire.

Thank you, R.D. Buell, Director of the Klamath County Walker Range Fire Patrol Association.

R.D.'s dedication over the decades has made communities safer and more resilient to wildfire.

Respect Pete Caligiuri, Forest Ecologist from the Nature Conservancy.

Relish the Deschutes Collaborative Forest Project for allowing me to participate as a guest.

Respect Chris Johnson, VP Operations Forester at Shanda Asset Management, LLC.

Chris shares his insight and is an original participant in the Deschutes Collaborative Forest Project.

Chris calls for less collaborative criticizing when implementing fire resilience.

Many thanks to Dr. Dan Leavell, Ph.D. of Forestry at Oregon State University (OSU).

Dan's faith in me to write, Axe-It-First shows his passion for education and immediate action.

Appreciate Dan and others at the Klamath Basin Research and Extension Center for forging partnerships in Firewise Communities, and Private Forests toward wildfire resilience.

Admire Oregon US Senator, Jeff Merkley, for his efforts and listening to my concerns.

Respect, Nick Smith, Founder of Healthy Forests Communities in 2013.

Nick's passion is for natural resources and rural communities. Nick worked in the Oregon Legislature and previously served as Chair of the Oregon Society of American Foresters Communication Committee and holds a BA in Journalism and Master's in Public Administration.

Many thanks to David Stowe of the Bend, Oregon Chapter of the Sierra Club, and respected participant of the Deschutes Collaborative Forest Project and the

Central Oregon Shared Steward ship Alliance. David shares a no nonsense, respected view.

David witnessed the San Diego, California's 2003 Cedar Wildfire that burned 273,246 acres at the rate of 3,600 acres per hour or 60 acres per minute. The fire incinerated 2,232 homes. Many of those homes were rated to be fire resistant with stucco siding and tile roofs but were burned to the ground. David embraces logging and prescribe burning on dry forest types.

Admire Homeowner Insurance Broker Sandra, Owner of Sandra Steury Insurance Agency, Inc.

Respect the Ph.D.'s, researchers and specialists who dedicate their lives to forestry and ecology.

Admire John Williams, Owner, and Founder of Quicksilver in LaPine, Oregon. Since 1983,

John dedicated his logging and forest restoration experience as the key to wildfire resilience.

About the Author:

Ron, standing next to a giant redwood in Jedediah Smith Redwoods State Park in 2002

Ron Rommel is an avid outdoor enthusiast, who, at an early age, learned to fish, horseback ride, mountaineered and later became a forest technician at the USFS from 1979-1984.

Served the Clackamas and Estacada Ranger Districts, on the Mt. Hood National Forest (NF).

Ron gained experience from Planning, Presale, Silviculture to Fire Prevention and Fire Management and earned a forestry degree in 1980. He became a member of the Society of American Foresters. Ron's education and exceptional experience led to his pragmatic insight of the forest. Much thanks to his Forestry Professors, the late Otto Olson and Dr. John Stuart, Forester Ph.D.

Later, Ron entered the private sector to become a Senior Certified Third-Party Log Scaler.

Ron's mentors included Master Check Log Scalers and Trainers Carl Everett and Glenn Crabb.

Many thanks for their insight into the Art of Log Scaling & Grading.

Ron scaled millions of board feet (BF) with westside and eastside scale.

Ron founded R. Rommel's Forestry Services which included clients: Weyerhaeuser, Olympic Resources, USFS and private timberland owners.

Ron inventoried thousands of acres of forest plant communities and commercial timberland.

As a contractor, Ron brought forth senior level expertise to field sampling of conifer, deciduous, herbaceous and vegetation inventories utilizing fixed and variable plot sampling methods.

Ron timber cruised millions of BF with boots on the ground across the Pacific Northwest (PNW).

Provided Weyerhaeuser and the USFS with reliable audits through accurate check cruising data.

Ron added a business degree to his experience in 1998.

Preface:

I grew up in the 1950's and 1960's surrounded by groves of large Douglas-fir near my family's Portland, Oregon home. The beauty of those majestic trees stuck with me defining my view of the forest. It brought me perspective that framed my life and respect for our Natural World.

Throughout the following decades, my frequent travels throughout Western and Central Oregon continued to grow within professional forestry. Somehow, I always felt the sins of the past would one day haunt us. That day has arrived for our forests and grasslands.

Since 2019, I served as a Fire Lookout at the Bald Mountain Fire Lookout Tower.

Bald Mountain is located, in Central Oregon on the Fremont-Winema National Forest.

My view from 7,400 feet above sea level often expands beyond thirty miles in all directions.

It is this view, which brings perspective to today's forest and the elements of Nature that affect it.

The 2020 wildfire season was devastating to Oregon. It was a perfect storm of events.

Wildfires are 85% -90% human caused, and early fires were found to be from arson ignitions.

As the hot weather developed and humidity levels dropped, lightning ignited more fires.

Between the arsons, unattended debris piles, abandoned campfires, downed powerlines, and lightning there was no shortage of ignition sources and plenty of federal forest to burn.

By 2021, sickened by the fact that our federal forest lands continued to be neglected I watched as forests grew increasingly susceptible to catastrophic wildfires. As a retired forestry professional, I felt compelled to act. After numerous conversations with others, I realized the events from the 1990's led to contentious attitudes toward federal management.

I reached out to Dr. Dan Leavell, Ph.D., Professor of Forestry at Oregon State University.

Dan's professional forestry experience spans from the early 1970's. Dan shared with me past events that shaped his life, including his actions that helped to save homes and people's lives.

It was his actions that encouraged me to share this vital information with others.

As drought continues to grip Oregon and the U.S. Western States, wildfire will be eminent.

Climate change requires heroic action to implement a plan to adapt forests and grasslands.

Time is of the essence!

We must embrace an enthusiastic study of today's forests and our need for immediate action.

Table of Contents

About the Author:	7
Preface:	9
Catastrophic-Mega-Wildfires Are Burning Every Year	13
Chapter 1	25
Chapter 2	29
Chapter 3	33
Chapter 4	39
Chapter 5	43
Chapter 6	51
Chapter 7	63
Build Quality Assurance into Forest Management	75
Homeowner's Evacuation Check Off List	77
Conclusion	80
Glossary	82
Bibliography	83

Catastrophic-Mega-Wildfires Are Burning Every Year

Today's forests are ready to burn and destroy human infrastructure and wild habitat communities. For decades, US Forest Service Forest (USFS) Policy reduced timber harvesting and suppressed wildfires. Lack of harvesting and severe fire suppression has shaped today's forests into tinder boxes ready for ignition.

"Most recently, on July 6, 2021, the Bootleg Fire, within the Fremont-Winema National Forest ignited from lightning and quickly grew to become Oregon's third largest fire since the year 1900. It destroyed 408 buildings, including: 161 houses, 247 outbuildings, along with 342 vehicles while venting a smoky haze stretching to the east coast of the USA. It burned 413,765 acres or 647 square miles and required 2,200 firefighters and multiple resources to get it 100% contained by August 15th." (2021 Bootleg Wildfire near Beatty, Oregon 2021, From Wikipedia 2021, 1)

Year 2021 Bootleg Wildfire, Oregon's 3rd largest since 1900

(Katu.com 2021) 2021 Bootleg fire's
HOT BURN

(New York Times 2021)
2021 Bootleg fire aftermath
Unmanaged BURNED, Managed SAVED

I personally interviewed Oregon Department of Forestry (ODF), Forester Randy Baley.

Randy's forestry experience spans 33 years and has firsthand knowledge of local forest lands. Since year 2014, along with follow-up observations in March 2021, Randy predicted the area prior to the Bootleg fire, which continuously experiences ongoing drought conditions, would be at the right time be an explosive fire challenge. (Baley 2022)

Randy was correct, the area he referenced blew up into Oregon's 3rd largest fire since 1900.

For all who studied and monitored the USFS Fremont-Winema National Forest property before and after the Bootleg fire, know that forest fuels reduction reduces catastrophic destruction.

Pete Caligiuri, Forest Ecologist from the Nature Conservancy indicated to me that there can be mixed results when only ladder fuels* are removed, and ground fuels are left in place on the landscape. (Caligiuri 2022)

*Ladder fuel is a firefighting term used to describe live or dead vegetation allowing a fire to climb up a tree canopy or hillside. Typical ladder fuels include tree branches, shrubs, and tall grasses both living and dead. They fuel the fires from the ground up to the largest trees in a forest. A fuel break is horizontal and vertical spacing between vegetation (live or dead) or other flammable materials reducing risk of fire's ability to spread. (Wikipedia, 2022)

Extreme drought impacts Oregon and throughout the Western States of the USA and the World. 21st Century catastrophic wildfires will impact wildland-urban-interface (WUI) communities.

Ron Rommel

2020 CATASTROPHIC-MEGA-WILDFIRES
HORRIFIC Catastrophic Wildfires Igniting Year after Year.

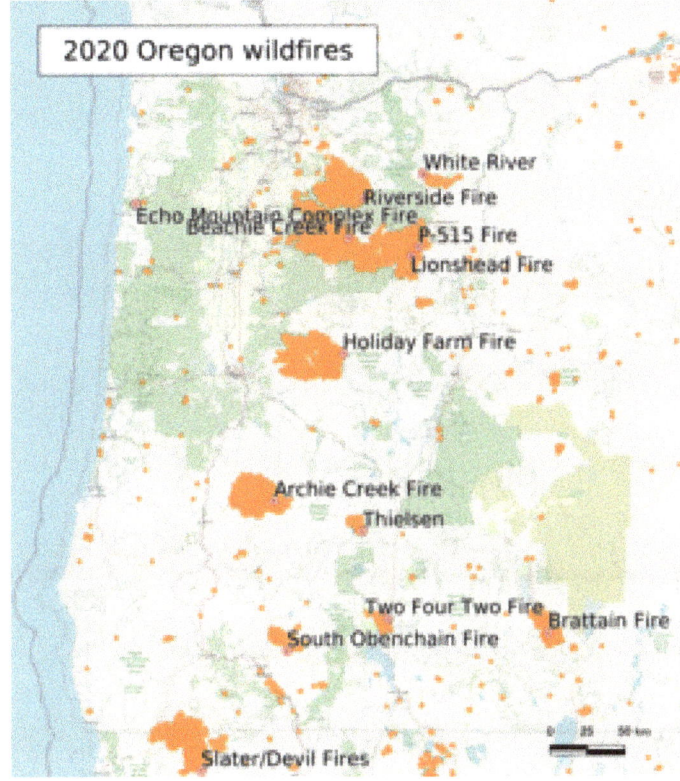

(en.m.wikipedia.org 2020)

Wildfires Ignite Year after Year

"The 2020 Oregon wildfire season was one of the most destructive on record in the state of Oregon. The season is a part of the 2020 Western United States wildfire season. Through the end of July 2020, 90% of Oregon's wildfires had been caused by humans. The fires killed at least eleven people, burned more than one million acres (400,000 ha) of land, and destroyed thousands of homes." (en.m.wikipedia.org 2020)

2020 fires burned an area the size of the entire Mt. Hood National Forest

Axe-It-First

"The Santiam Fire was an extremely large wildfire that burned in Marion, Jefferson, Linn, and Clackamas Counties in Northwest Oregon, United States. Having ignited in August of 2020, the 402,274-acre (162,795 ha) fire ravaged multiple communities in northwestern Oregon before it was fully contained on December 10, 2020.

The fire started as three separate fires. The Beachie Creek, Lionshead, and P-515 fires were ignited by lightning on August 16, 2020. The first three fires gradually grew, before explosively spreading in early September during a heatwave, fanned by powerful east winds. Early on September 8, the Beachie Creek and Lionshead Fires merged, and the combined fire was labeled the Santiam Fire, before being returned to their original names a couple of days later.[7]

The P-515 Fire merged into the Lionshead Fire a few days later."[8][7] (en.m.wikipedia.org 2020)

"On the morning of August 16, thunderstorms moved across Oregon, starting multiple fires, including the Beachie Creek Fire, the Lionshead Fire, and the P-515 Fire. The Lionshead and P-515 Fires were ignited in the Warm Springs Indian Reservation, near Mount Jefferson, while the Beachie Creek Fire was ignited near Opal Creek." (en.m.wikipedia.org 2020)

Working from the Bald Mountain Fire Lookout, I heard Wildfire Response Team's radios for 2 weeks without extinguishing the Lionshead and P-515 fires. On September 7, 2020, a fierce 50+ mph east wind blew the Lionshead and P-515 into Western Oregon.

Massive Catastrophic Destruction Ignites Western Oregon

Multiple events followed the east wind and contributed to further ignition sources. Those events included: abandoned campfires, downed powerlines, dry forest fuels and low humidity fire weather conditions, bringing about the most destructive fire season in Oregon.

2020 Riverside Wildfire Grows to 112,000 Acres (Katu.com 2021)

Riverside Wildfire grew to a total of 138,054 Acres (en.m.wikipedia.org 2020)

2020 Holiday Farm fire destroys Blue River home   2020 Santiam Complex Wildfires

Much of the town of Blue River was destroyed   Gates Elementary School destroyed

Talent, Oregon September 9, 2020,   Oregon Coast near Otis (OregonLive.com 2020)

These few photos define the aggressive and horrific degree of destruction in Oregon.

(en.m.wikipedia.org 2020) (Photos accessed by Author March 2022)

Oregon's catastrophic-mega-wildfires has impacted the Insurance Industry, Farmers and Safeco refuse to write new homeowner policies in Oregon Firewise Communities. ((Steury) 2022)

2003 Devastation to the State of California

Santa Ana East Wind Blasted a Human Ignition and Torched San Diego

CEDAR FIRE BURNED ALL THE WAY TO THE PACIFIC OCEAN

(Wikipedia 2003)

Investigation

"Investigators determined that the fire was started by Sergio Martinez of West Covina, California, a novice hunter who had been hunting in the area and had become lost.[21] Martinez initially told investigators that he had fired a shot from his rifle to draw attention and that the shot had caused the fire,[22] but he later recanted and admitted he started the fire intentionally to signal rescuers. After gathering sticks and brush together, Martinez lit the brush and quickly lost control of the fire because of the heat, low humidity, and low moisture content of the surrounding vegetation." (Wikipedia 2003)

David Stowe, of the Bend, Oregon, Chapter of the Sierra Club, and respected participant of the Deschutes Collaborative Forest Project and the Central Oregon Shared Stewardship Alliance, witnessed San Diego, California's 2003 Cedar Wildfire that burned "… 273,246 acres at the rate of 3600 acres per hour." "…destroyed…2,232 homes and killed fifteen people, including one firefighter." (Wikipedia 2003)

"Many of those homes were rated as fire resistant with stucco siding and tile roofs but were burned to the ground." (Stowe 2022)

David sees the need for managing both forestlands and grasslands to restore wildfire resilience. (Stowe 2022)

Axe-It-First

"The fire remains one of the largest wildfires in California history and, as of 2020,[4] …According to CALFIRE, it is also the fifth deadliest and fourth-most destructive wildfire in state history,[5][6] causing just over $1.3 billion in damages. [1]" (Wikipedia 2003)

If this can happen in San Diego, it can happen in your community!

2002 Missionary Ridge Fire Ignites Near Durango, Colorado Missionary Ridge Wildfire

(the-Journal.com 2002)

(Library.org 2002)

"The Missionary Ridge Fire began on June 9, 2002, northeast of Durango in southwest Colorado. It burned until July 15, destroying forty-six houses and cabins and charring 73,000 acres of La Plata County Forest. One firefighter died while fighting the blaze, which became the seventh-largest wildfire in Colorado history." (Encyclopedia.org 2002)

"In the spring of 2002, Colorado's mountain snowpack stood at just 53 percent of its average, and a warm, dry spell in April and May melted away all that moisture before the summer heat arrived in June." (Encyclopedia.org 2002)

During the Missionary Ridge Fire, Dr. Dan Leavell was the acting Type 1 Operations Section Chief. Every evening downdraft winds drove fires to burn structures. When the Chief Mountain Hotshots completed burn outs at the head of the fire, Dan released Air Attack. (Dr. Dan Leavell 2022)

Dan said, "I was walking down – across the main river and outside of our fire perimeter, where there were many homes in the path of a new fire started by a figure too far away to discern whether male or female. The person ignited the fire and ran within vegetation and hid from view. The wind, which had picked up to 30-40 mph, immediately grabbed the flames and torched a flaming front over one hundred feet high toward a community of homes in its path. Later investigation revealed the fire started when that individual shorted out an electric wire fence." (Dr. Dan Leavell 2022)

Dan said, "I had two seconds to decide to deploy the thirteen aircraft just released to save the homes – or not. The fire started was outside our fire perimeter and therefore outside our jurisdiction. My Air Attack friend called me on the radio at second two: 'What do I do, Chief?' On the third second, my answer was, "Get everyone together and beat the hell out of it."

(Dr. Dan Leavell 2022)

Since it was outside of Dan's authority, the incident commander revoked Dan's title. Dan said, "The following day, I decided to stop by the Falls Creek Ranch to make sense out of this momentous change in my life. Driving by the burnt homes and homes with black ten feet from decks – but saved. Dan saw a group of people crying, laughing, introduced himself and asked how they were doing. They were so grateful for the fire resources that saved much of their homes, property–and most importantly–their lives." (Dr. Dan Leavell 2022)

"Today, the neighborhood devotes three days a year to fire-mitigation work, such as thinning, trimming, or clearing brush."

This case study was presented by Amanda Brenner who was hired by our colleague, Glenn Ahrens (OSU Extension Forest Agent for Clackamas, Hood River, and Marion County), to assist managing the Clackamas Tree School. (Dr. Dan Leavell 2022)

Chapter 1

Why Are There So Many Catastrophic-Mega-Wildfires?

For Oregon, the short answer is the lack of U.S. Forest Service (USFS) management. The USFS manages nearly 50% of all forestlands in the State of Oregon. Before 1990, the USFS annually sold an excess of ten billion Scribner Board Feet of timber from Oregon. Human desire to liquidate old-growth fire resilient forests led to the fight. The early 1990's environmental lawsuits triggered the lack of USFS management. That lack of management impeded three decades of robust harvest and degraded trust between the environmental community and the timber industry on federal lands. Since most wildfires are human caused, there remains a huge risk for ignition on USFS lands.

Catastrophic Wildfires Billow Volumes of Smoke That Threatens Life
Smoke exasperates climate change, compromises the public, and kills the most susceptible.

AQI Category and Color	Index Value	Description of Air Quality
Good — Green	0 to 50	Air quality is satisfactory, and air pollution poses little or no risk.
Moderate — Yellow	51 to 100	Air quality is acceptable. However, there may be a risk for some people, particularly those who are unusually sensitive to air pollution.
Unhealthy for Sensitive Groups — Orange	101 to 150	Members of sensitive groups may experience health effects. The general public is less likely to be affected.
Unhealthy — Red	151 to 200	Some members of the general public may experience health effects; members of sensitive groups may experience more serious health effects.
Very Unhealthy — Purple	201 to 300	Health alert: The risk of health effects is increased for everyone.
Hazardous — Maroon	301 and higher	Health warning of emergency conditions: everyone is more likely to be affected.

(https://www.epa.gov/wildfire-smoke-course/wildfire-smoke-and-your-patients-health-air-quality-index)

(EPA and Agency n.d.)

Catastrophic-mega-wildfire smoke reaches Unhealthy, Very Unhealthy and Hazardous levels.

Serving as a Fire Lookout, I personally witnessed the need to wear an industrial gas mask. When you find you can only breathe with a gas mask, you know conditions are **HAZARDOUS**.

We can agree, uncontrolled wildfires kill people and wildlife and destroy the human infrastructure of cities, towns, communities, businesses and devastate wildlife habitat and watersheds.

"The main effect of wildfires on our wildlife is immediate loss of habitat and a reorganization of animal communities." (Oregon Department of Fish & Wildlife 2020)

"If we fail to act, there will be little left to preserve." Ron Rommel.

Research Speaks to All Who Will Listen

Congressional Research Service Reveals Wildfire on Federal Land Doubled Since 2016, "Of the federal acreage burned nationwide in 2020, 68% (4.8 million acres) burned on FS land and 32% (2.3 million acres) burned on DOI land (see Figure 3). Most wildfires are human caused (88% on average from 2016 to 2020), although the wildfires caused by lightning tend to be slightly larger and burn more acreage (55% of the average acreage burned from 2016 to 2020 was ignited by lightning)." (Congressional Research Service 2021)

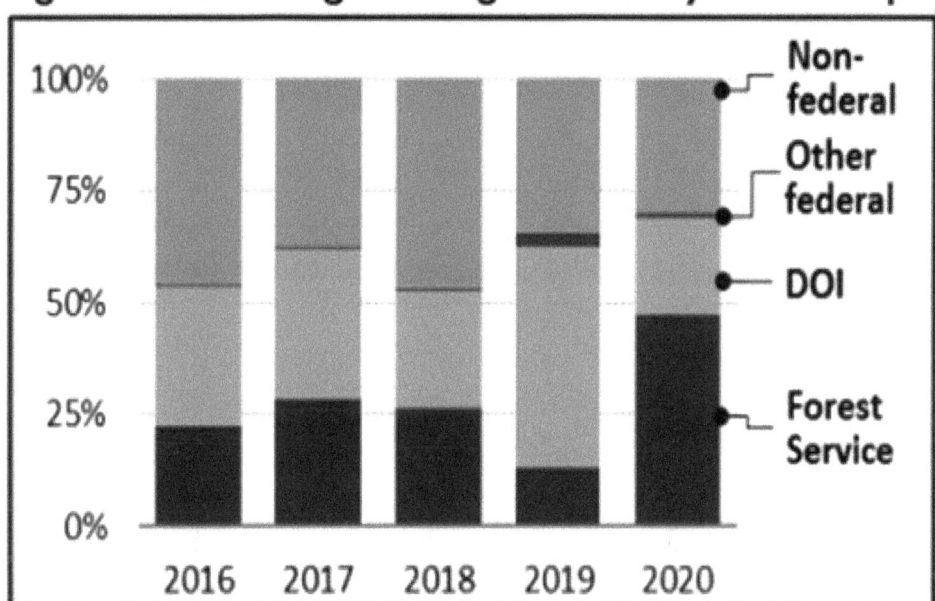

Figure 3. Percentage Acreage Burned by Ownership

Source: NICC Wildland Fire Summary and Statistics annual reports.

Figure 4. Acreage Burned by Region and Ownership

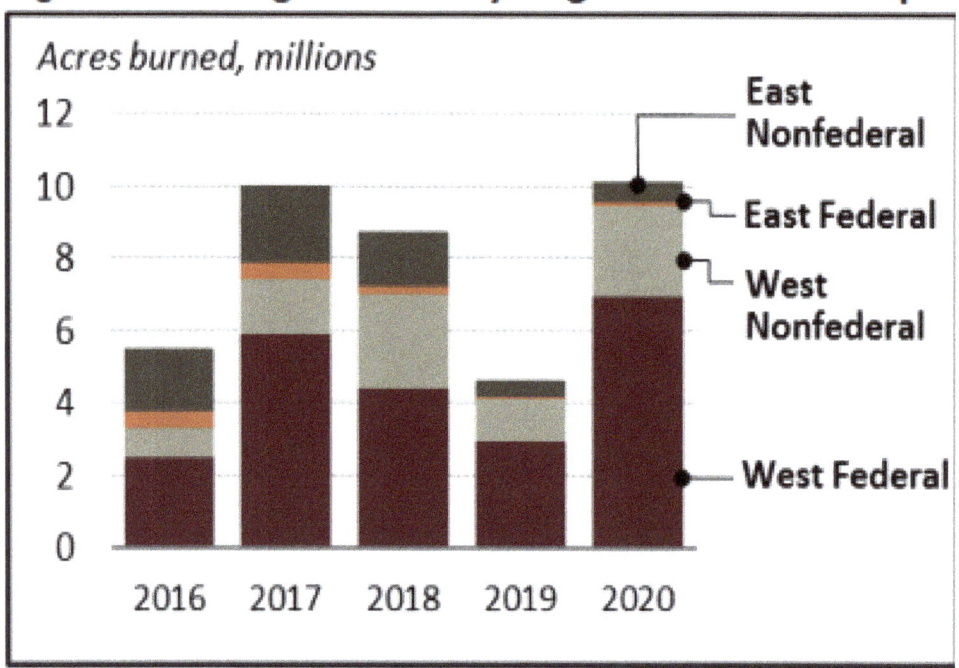

Source: NICC Wildland Fire Summary and Statistics annual reports.

"In the West, most of the fires occur on federal lands (see Figure 4). In 2020, 81% (0.5 million acres) of the acreage burned in the East was on nonfederal land, whereas 75% (7.1 million acres) of the acreage burned in the West was on federal land." (Congressional Research Service 2021)

This congressional research supports the need to reduce forest fuels on federal timberlands.

Climate change is upon us for decades to centuries. We Must Act!

Chapter 2
How Did Wildfires Become Catastrophic-Mega-Wildfires?

Virgin forests minimized catastrophic wildfires.

Humankind's lack of foresight promoted the opposite.

Today's wildfires became catastrophic-mega-wildfires from a lack of the GIANTS and the accumulation of high-density forest fuels and severe drought triggered by climate change. A lack of soil moisture from snow melts and precipitation, coupled with solar heating and low humidity, dries out the forests to a point, at which, they are ripe for wildfire. Today's forests are extremely prone to wildfire due to their composition and densities, they are a tinderbox ready to explode from ignition during fire weather.

Early historical human deforestation led us to our current situation with few remaining giant thick bark, fire resistant conifer trees to withstand catastrophic wildfires.

Loggers in Clatsop County, Courtesy Ore. Hist. Soc. Research Lib., OrHi93132
Early Oregon Logging, Date is unknown, oregonencyclopedia.org (Robbins unknown)

Today's eastside forest is a vivid contrast from the early virgin timber

Overcrowded forests yield High-Density Forest ladder fuels (Author's Photo)

Wildfire Can Ignite into MASSIVE IGNITION On the Larger Landscape

Fire Impacts the Landscape: Let us take a moment to review the components of FIRE.

Fire requires three components to ignite. Oxygen, Fuel, and Heat comprises the fire triangle.

Removing just one of the components prevents a fire.

In a forest, the only controllable component during a wildfire is FUEL!

The reduction of forest fuel is essential to control a wildfire

(Kathy Sarns, Alaska, Artist's rendition of the Fire Triangle – Source: USFWS 2006)

Radiant Heat Transfer triggers fire behavior. Topography and aspect influence this behavior.

Radiant Heat Transfer spreads a fire as the fire front moves closer, and radiant heat increases. (unknown, Radiant Heat Transfer unknown)

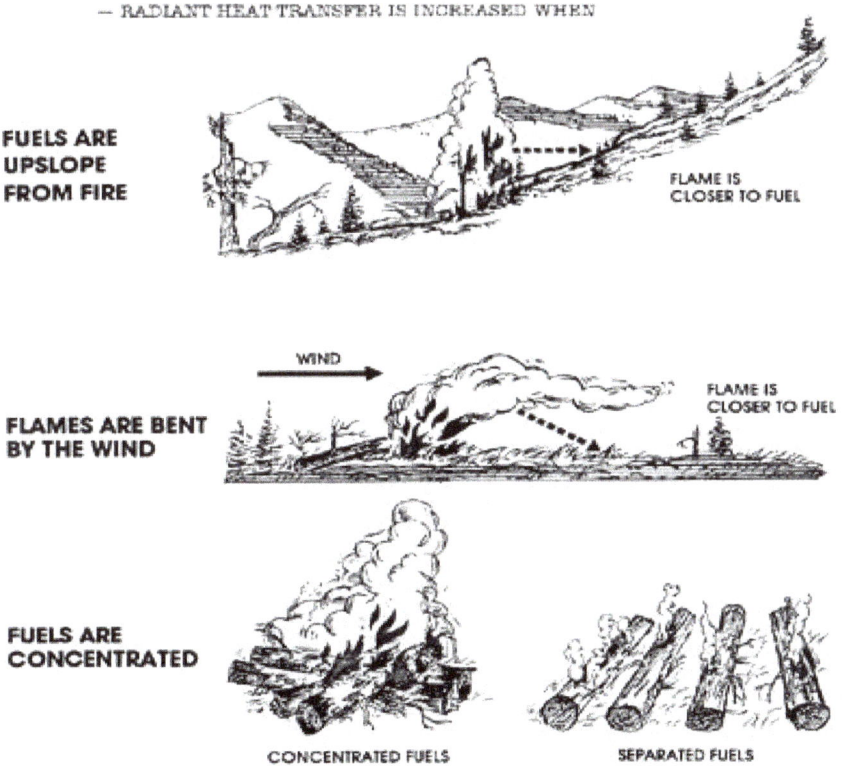

Chapter 3
History Shaped Today's Wildfires

Nature has shaped the forest over eons of time. Humans accelerated deforestation.

Giant Douglas-fir growing on Mt. Scott, Portland, Oregon
8.5 feet in diameter & 300 feet tall. Oregonian May 26, 1912
(Oregonian 1912) Author accessed image and photo January 2022

Historically, "By the early Pleistocene, some 1.5 million years ago, and before major glaciation, the flora of the Pacific Northwest was essentially established as it appears today." (Franklin 1979)

Nature is our greatest teacher. Prior to the European settlement of North America, nature was shaping our forests with fire. Over the centuries, nature slowly grew the GIANT trees to become resilient to wildfires.

Over eons of time, virgin forests grew, died, and evolved to become fire resilient giants. Early Indigenous native peoples shaped their habitats with fire. They learned from nature.

Early peoples witnessed lightning and volcanic eruptions making fire. Fire became their tool. They burned to encourage native forbs and grasses to bring wild game to their habitat. This meant meat for their people. Those habitats sustained the early tribes.

In vivid contrast, "New England Puritans believed that the wilderness was the natural habitat of the devil. Since Native Americans belonged to the wilderness, their familiarity with the ways of the devil seemed obvious to the settlers." (Text by Aileen Agnew, 2021 17th Century)

American expansion led to rapid deforestation.

Historical conquest of America led a European desire to convert the forests into communities, towns, and cities due to the common belief that forests were endless. Then, forward thinking people spoke.

"A new era of forestry began in 1905 as Congress transferred the Forest Reserves from the Department of the Interior to the Department of Agriculture. The close warm relationship and respect between President Theodore Roosevelt and the Nation's Chief Forester Gifford Pinchot fostered the development of resource conservation." (William W. Bergoffen 1976)

Aldo Leopold once wrote, "It is inconceivable to me that an ethical relation to land can exist without love, respect, and admiration for land, and a high regard for its value." (Leopold 1949)

Walk-Through History and the Timber Wars of the 1990's

My love for trees began when I grew up among groves of Douglas-fir near my family home.

I enjoyed family summer trips to Central Oregon viewing yellow Old-Growth Ponderosa pine. As a child of the 1950's, I first heard about wildfire from my grandfather, Fred. He shared with me the devastation following the aftermath of the 1933 Tillamook Burn.

Later as a US Forest Service (USFS) Forestry Technician, from the late 1970's through the early 1980's, I overheard attitudes that advocated liquidating old growth by cutting it back and burning it black. I did not embrace that attitude; I witnessed the harvesting of old-growth timber. Many of those giants were centuries old and were compromised with broken tops and advanced decay. Many of the large old trees were in a state of decline, yet there were many that could yield sound wood and may have lived longer.

Since the timber industry wood market remained strong and demand was high, those large old trees made structurally strong lumber from their stable and fine grain composition.

It was my hope industry and conservation would find a balance.

Later, as a Senior Third-Party Log Scaler, I lived through the timber wars of the early 1990's. I witnessed daily battles between the environmentalists and the timber industry. Both sides echoed their points of view. Environmentalists waged war with industry. When industry ignored the environmentalist view, the battle grew larger, and lawsuits followed. Most federal timber sales ended up in court and were blocked from further harvest. On private forestlands, the accountant yielded greater power than the forester.

In 1992, President Bill Clinton supported a new Northwest Forest Plan. It echoed ecology to save Old-Growth forests. It was a plan to save habitat for the Northern Spotted Owl. It devastated loggers and the timber industry. Hundreds of log scalers faced layoff and job loss, and 6,000 USFS timber employees lost their jobs through either attrition or retirement. Severe curtailed timber sales caused a loss of jobs within the USFS and the timber industry. It was a double-edged sword between balancing the needs of people with ecology.

Today's forests are at a dire turning point, as to whether we, as humans, have the capacity to right the wrongs of the past and begin to embrace a new management paradigm of forestry. Climate change is a game changer for forest survival. We must come together as natural resource professionals, environmentalists, Indigenous tribes, timber industry, communities, and stakeholders. By coming together, we will all benefit from a heroic attempt to help forests, grasslands, inclusive habitats, and watersheds to adapt to climate change as we attempt to protect human infrastructure and build wildfire resilience. We faced a critical time and those moments in history led legislation to effect change.

The difference between today and the past is infrastructure.

Human Infrastructure dominates the landscape and grows closer to forest and grasslands.

The Catalyst for Change Began with 21st Century Federal legislation:

"Healthy Forests Restoration Act: The Healthy Forests Restoration Act of 2003 significantly changed the way the Forest Service and the Bureau of Land Management do fuel reduction projects to reduce the risk of catastrophic fire. In addition to specifying how special fuels reduction projects should be created and implemented, the law also promotes use of biomass and small diameter materials; creates a forest reserve program; provides technical assistance for private landowners and addresses insect infestations and other environmental threats to healthy forests."

(USDA Department of Agriculture, Forest Service unknown)

"Stewardship Contracting: The 2003 Appropriations Act (16 U.S.C. 2104 Note) provides the Forest Service and the Bureau of Land Management ten-year authority to enter into stewardship contracts and agreements. Stewardship contracting is intended to promote collaborative working relationships with local communities, improve land conditions, and help develop sustainable rural economies by developing, implementing, and monitoring projects collaboratively. Final agency direction became effective December 12, 2005."

(USDA Department of Agriculture, Forest Service unknown)

Beginnings of Collaboration and the Law-

"The Forest Service and other federal agencies are authorized to work collaboratively with the public under a variety of laws and directives. Empowered by such, we collaborate regularly with partners, including tribes, states, other federal agencies, nonprofits, businesses, and communities." (USDA Department of Agriculture, Forest Service unknown)

"Collaborative Forest Landscape Restoration Act: Congress, under Title IV of Omnibus Public Land Management Act of 2009 (PDF, 40 KB), established the Collaborative Forest Landscape Restoration Program (CFLRP). The purpose of the CFLR Program is to encourage the collaborative, science-based ecosystem restoration of priority forest landscapes." (USDA Department of Agriculture, Forest Service unknown)

As the Deschutes Forest Collaborative Project has pointed out, it is time "…to put mistrust, and past conflict behind us-and pick up the quality, pace, and scale of forest restoration projects in the dry fire-adapted forests of Central Oregon." (Deschutes Collaborative Forest Project unknown)

Partnerships or Collaboratives?

What is the best approach?

As stated, "The Forest Service and other federal agencies are authorized to work collaboratively with the public under a variety of laws and directives. Empowered by such, we collaborate regularly with partners, including tribes, states, other federal agencies, nonprofits, businesses, and communities." (USDA Department of Agriculture, Forest Service unknown)

A true partnership must show trust, be collaborative, and build solidarity. A collaborative without partnerships is like trying to fight fire without water and resources. Working together, is effective for habitats and the WUI to adapt to climate change.

Dr. Dan Leavell Ph.D. of Forestry at Oregon State University (OSU) whole heartedly defends partnerships as the key to promote and expand, science based, fire resilience on government, tribal, private forestlands, and WUI. Cooperation will make the difference. (Dr. Dan Leavell 2022)

"FACILITATING RESTORATION PROJECTS ON PUBLIC AND PRIVATE FORESTLAND IN KLAMATH AND LAKE COUNTIES THROUGH EDUCATION, OUTREACH, AND DIVERSE PARTNERSHIPS."

(Klamath Lake Forest Health Partnership (KLFHP) 2020)

KLFHP is committed to:

- "Providing technological and ecological information on forest health;"
- "Serving as a resource for all forest landowners in diagnosing and addressing forest health problems;"
- "Working cooperatively with landowners, the general public, and forest operators to educate and encourage best management practices on forest lands; and"
- "Using innovative partnerships and funding sources to increase the pace and scale of Restoration across public and private lands." (Klamath Lake Forest Health Partnership (KLFHP) 2020)

Chris Johnson, VP Operations Forester at Shanda Asset Management and one of the original participants of the Deschutes Forest Collaborative Project calls for less collaborative criticism for fire resilience. (a. o. Chris Johnson 2022)

As we seek collaboration and form partnerships, **Time is of Essence.** We need heroic action.

"Never doubt that a small group of thoughtful, committed, citizens can change the world. Indeed, it is the only thing that ever has."

Margaret Mead, Anthropologist, Ph D. (Margaret Mead 1978)

Chapter 4
The Effects of Climate Change in Todays Forests

As a child of the 1950's growing up in western Oregon, I helped my dad care for horses.

We always waited after July 4th to load hay bales from the fields, because before then, it was too wet. I witnessed an annual snowpack on Mt. Hood at Timberline Lodge at 20 feet.

Today it is barely ten feet. I point out this fact that our climate has dramatically changed.

Since 1980 climate has steadily warmed and today there is no question of its impact on forests.

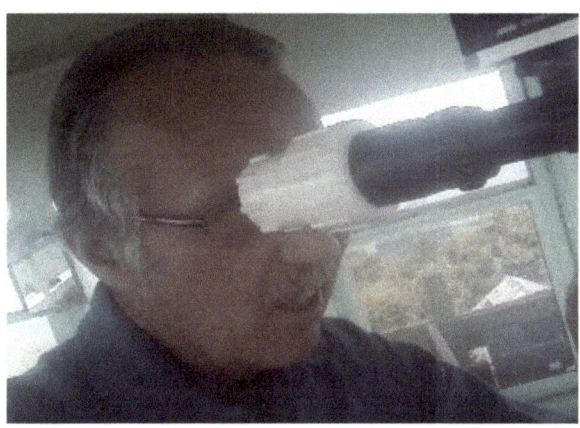

Ron on Bald Mtn. Lookout From year 2019

Darlene Wildfire LaPine, OR 2021

Bald Mtn. Tower Fremont-Winema

Axe-It-First

Staffing the Bald Mountain Fire Lookout on the Fremont-Winema National Forest, I know firsthand, the dire situation of our National Forests and our recent catastrophic-mega-wildfires. At this very moment, Oregon's east-westside forests are facing a severe cataclysmic problem. We must adapt forest and grasslands to climate change.

Scientific Consensus: Earth's Climate Is Warming

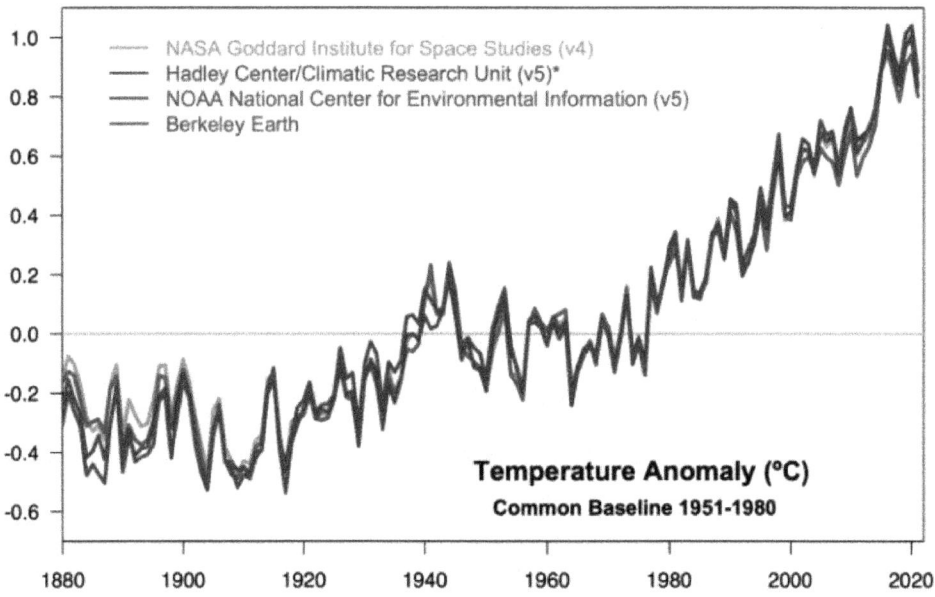

(Scientific Consensus: Earth's Climate Is Warming 2020)

Whether climate change remains beyond decades or centuries,

"We need to Act Today."

Recent Research Maps Low to High Burn Probability Areas in Oregon-

In a recent study by PLOS ONE, the following diagram and article was released: "The intensity and scale of wildfires has increased throughout the Pacific Northwest in recent decades, especially within the last decade, destroying vast amounts of valuable resources and assets. This trend is predicted to remain or even magnify due to climate change, growing population, increased housing density. Furthermore, the associated stress of prolonged droughts and change in land cover/land use puts more population at risk…"

(PLOS ONE | https://journals.plos.org/plosone/article?id=10.1371/journal.pone.0264826 (March 8, 2022)

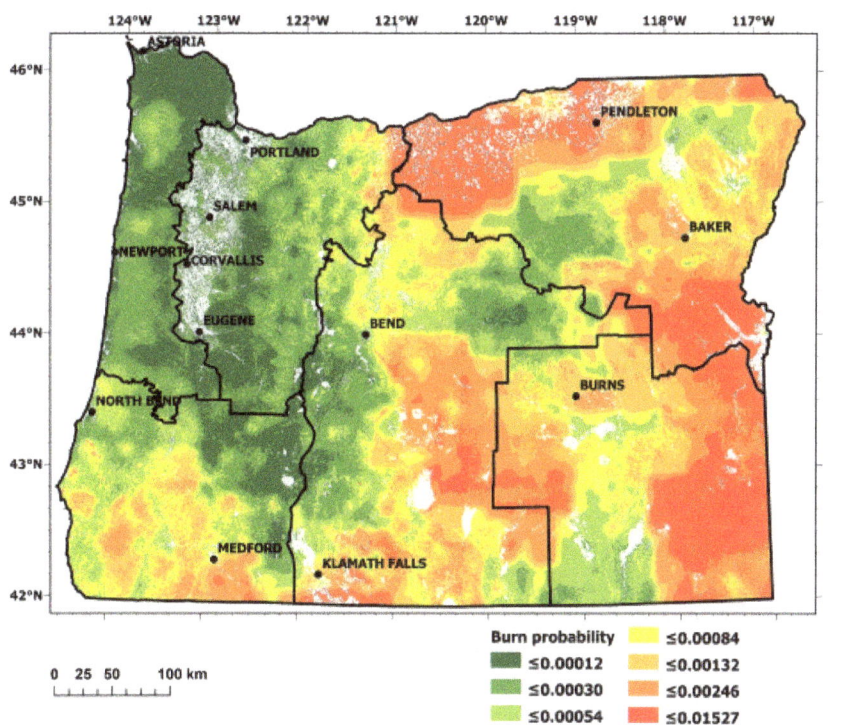

Fig 3. Burn probability for all FSAs using worst-case meteorological data on record during peak fire season from the 50 selected weather stations.
https://doi.org/10.1371/journal.pone.0264826.g003

Burn probability risk is colored as; dark green being the lowest and dark orange the highest.

Schmidt A, Leavell D, Punches J, Rocha Ibarra MA, Kagan JS, Creutzburg M, et al. (2022) A quantitative wildfire risk assessment using a modular approach of geostatistical clustering and regionally distinct valuations of assets—A case study in Oregon. PLoS ONE 17(3): e0264826. https://doi.org/10.1371/journal.pone.0264826

Axe-It-First

Chapter 5

Adapting Forests and Grasslands to Climate Change

A Look at Western Oregon Forests as seen through My Eyes

Westside forests from the Oregon Coast to the western crest of the Cascade Mountains typically receive anywhere from 80+ inches of precipitation on the coast to 60+ inches near the crest of the Cascades. These forests benefit from both precipitation and much older soils.

Some soils are 100,000 years old. "Until about twelve million years ago, Western Oregon was on the floor of the Pacific Ocean…" (Geology, Soils & Climate unknown)

Yet, Eastside Klamath County, soil is younger from the eruption of Mt. Mazama 7,800 years ago.

If it takes between 500 to 1000 years or more to form 1 inch of soil from rock, then we can extrapolate that it will take from 50,000 to 100,000+ years to form a 100-inch depth of fine particles, like clay soil. Dr. John Stuart, Ph.D. taught that climate, organisms, relief, and parent material over time builds soil. Dr. Stuart often referred to this process as: CLORPT.

So why is this important to mention?

Fine particle soils retain a higher moisture content that is beneficial to westside forests. Over eons of time, older soils accumulate organic matter from the decomposition of plants and older remnants of rotting trees and logs. The westside's wetter climate elevates soil moisture on forest growing sites. It is well known that north/north-east/northwest aspects tend to receive less solar radiation and retain higher soil moisture. Whereas, south/southwest/southeast aspects receive more solar radiation and retain less soil moisture, thus they become drier sites.

During my years of inventorying plant communities and timber cruising, it became obvious to me the driest sites are much more problematic when it comes to wildfire. As sites exist at various elevations of topography, it is the mid-slope that gets the warmest. Stream drainages are the least problematic when it comes to wildfires. When inventorying USFS forest plant communities and timber types along the Western Central Cascades within the moist

Western hemlock zone exhibiting well drained silty clay soil, the management implications reveal "Sites are resistant to the effects of moderate intensity fire."(Diaz 2002)

Moist sites are more difficult to ignite and more resistant to wildfire. These sites may not require frequent forest management treatment to reduce fuel loading.

Oregon's westside forests receive more moisture increasing decomposition of forest fuels. Oregon's eastside forests receive less moisture slowing decomposition to build forest fuels.

As drought continues to impact forests and grasslands, there will be less moisture retention.

Fauna and flora species will either adapt or die. Fuel load reduction will help our attempt to adapt our forests and grasslands to climate change.

Climate change requires intervention to adapt forest and grasslands

Deploying the know-how through knowledge

My business professor once said, knowledge is like a bag of tools. As a timber cruiser/inventory specialist working as a forest technician, forester, and log scaler, I utilized my tools, knowledge, and experience to routinely observe, quantify, and record forest data.

Ron Rommel

Authors photos:

Point sampling and Aging from USFS Oregon's Eastside Log Scaling, Cone Lumber Eugene, OR Timber check audit USFS Westside.

My experience has led me to this moment in time. Sharing a bit of my knowledge will help you begin the process of implementing a new paradigm for today's forests.

Creating forest resilience-

To achieve long-term forest resilience to catastrophic wildfires, the components of the forest must be in balance to promote a healthy forest ecosystem. Those components reflect the overall composition of the forests. Reducing high density stands of trees, ground, and ladder fuels, with a combination of mechanical removal and prescribed fire will enhance fire resilience.

Remember the Fire Triangle?

(Kathy Sarns, Alaska, Artist's rendition of the Fire Triangle – Source: USFWS 2006)

In a forest, the only controllable component during a wildfire is FUEL. The reduction of forest fuel is essential to control a wildfire.

Remember, Today's high-density forest fuels are stacked like a tinderbox ready for ignition.

When managing forests on a large landscape, it is vital to first develop a strategy that prioritizes the reduction of forest fuels in the closest proximity to human infrastructure.

Forest Hierarchy Through Silviculture

"Silviculture is defined as the art and science of controlling the establishment, growth, composition, health, and quality of forests and woodlands to meet the diverse needs and values of landowners and society on a sustainable basis." (Helms 1998, USDA Forest Service 2004) (David C. Powell 2013)

"In the early 1990's, the advent of ecosystem management created some confusion about silvicultural terminology and how it should be applied. Whether it was appropriate or not, some land managers were abandoning historical definitions and creating a new silvicultural term, even when the new terms were being used to describe old (traditional) ways of managing the forest." (David C. Powell 2013)

Axe-It-First will focus on size and hierarchy over the landscape which limits the frequency of catastrophic wildfires to once in a century or longer, rather than, occurring each year.

As a forester, I was taught the largest trees that overshadow the landscape are the Dominants. They occupy the most space and basal area per acre. In ecological terms, "Ecological dominance is the degree to which one or several species have a major influence controlling the other species in their ecological community due to their large size, population, productivity, or related factors or make up of more of the biomass". (Wikipedia.org 2021)

Why is this important?

(Author photos of Central Oregon forest, Winter 2022)

Dominant remnant ponderosa pine   CO-Dominant ponderosa pine   Intermediate ponderosa pine

Tree dominance, within a forest, impacts the competition for sunlight, moisture, and nutrients. The dominant trees limit sunlight, moisture, and nutrients for understory trees. When I walk through dense dominant stands of old growth displaying their majestic height and broad crowns where the lowest branches tower well above the forest floor, you soon realize the significance of dominance through the reduction of understory thickets yielding fire resilience. Today's high-density stands are subjected to stunted growth, insects, disease, and mortality. Dominance yields fire resilience.

Dominance yields fire resilience during drought conditions.

Those giant trees of yesteryear evolved massive crowns with the first branches extending from 100 to over 180 feet in Western Oregon (Oregonian 1912) above the forest floor.

Those first branches were resistant to ground fires. Even the virgin Ponderosa Pine were impressive. Many giants grew the thick fire-resistant bark so wildfire could not penetrate to their live wood. They lived for centuries, and many Douglas-fir lived beyond 1,000 years of age.

Recent research indicates large dominant trees have the capacity to store 42% of forest carbon. (David J. Mildrexler 2020)

Climate change may impact our Earth for decades or centuries, it is vital that humans attempt to right the wrongs of our past and begin an effort to manage forest dominance.

What do you keep and what do you cut?

Nature and forests are not static, but rather dynamic and are constantly in a state of life and death, routinely, managing this growth is key to preventing catastrophic-mega-wildfires.

Before cutting, preserve the dominate, co-dominate, intermediate, and saplings that are the phenotypic or best trees to propagate the future forest. Cut the diseased and insect infested high-density trees.

(Authors photos of Central Oregon forest, Spring 2022)

Phenotypical Ponderosa Pine
(BEST LOOKING & DISEASE FREE)

Lodgepole old sucker & canker

Search for the absolute best phenotypical tree that may be a dominant, co-dominant or intermediate. Co-dominant trees are just below the dominants and intermediates are just below co-dominants. The trees should be the healthiest and be free from disease and rot.

My forestry professor, Otto Olson, once said, "In the past they cut the high-grade and left the junk." High-grade logging from the past took the best trees and left the remainder. What humans deforested in the past took nature eons to create resilient forests. Today's forests need a helping hand, first through utilization followed by prescribed fire.

Chapter 6
Federal Forest Growth Threatens Wildland-Urban-Interface

Let us review a comprehensive 10-Year Forest Inventory- Abstract (FIA):

Palmer, Marin; Kuegler, Olaf; Christensen, Glenn, tech. eds. 2018. Oregon's forest resources, 2006–2015: Ten-year Forest Inventory and Analysis report. Gen. Tech. Rep. PNW-GTR-971. Portland, OR: U.S. Department of Agriculture, Forest Service, Pacific Northwest Research Station. 54 p.

"Oregon has thirty million forested acres that cover roughly half the state's land area. This report provides detailed estimates of forest area, tree species composition and distribution, volume, biomass, carbon, standing dead trees and down wood, and understory vegetation on forest land for the state of Oregon based on the annual FIA forest land inventory through 2015. It also includes the first estimates of annual growth, mortality, and removals on forest land available from remeasured annual inventory plots, representing 50 percent of the full 10-year cycle. The FIA program collected inventory data on 9,439 forested plots during the 2006–2015 measurement cycle. Three-fourths of this forest volume occurs on the moist west side of the state." (John Chase 2018)

"**Summary Key Forest Inventory and Analysis (FIA) Statistics, Oregon, 2006–2015**

- Number of forested plots measured by the FIA program (2006–2015): 9,439
- Estimated total forest area: 29.7 million ac
- Estimated number of live trees: 10.3 billion
- Estimated net live tree volume: 106.9 billion ft3
- Estimated aboveground net live biomass: 2.2 billion tons
- Estimated aboveground net live carbon: 975.6 million Mg" (John Chase 2018)

Let's put this into perspective-

Within Oregon's 30 million forested acres there are ten billion live trees that yield

107 billion ft3 or nearly, one billion metric tons of stored carbon, nearly 1 trillion Scribner board feet (BF). (Scribner Rule) (¾) of the total volume occurs on the westside of the Cascade Range that is dominated by Douglas-fir. (¼) of the total volume occurs on the eastside of the Cascade Range.

Forest composition-

The above report shows a preponderance of young tree age classes from 1 to 100 years of age. Much fewer acres of older tree age classes beyond 101 years of age. (John Chase 2018)

Forest productivity from the above report shows-

Western Oregon National Forests are producing 120 cuft/ac/yr or 1440 bf/ac/yr.

Eastern Oregon National Forests are producing between 20-50 cuft/ac/yr or 240-600 bf/ac/yr.

Abstract and further text is available for your review. (John Chase 2018) (See following Charts)

US Forest Service Reduction in Management Increases Mortality While Building Annual Growth for Westside-Eastside Forests

General Technical Report PNW-GTR-971

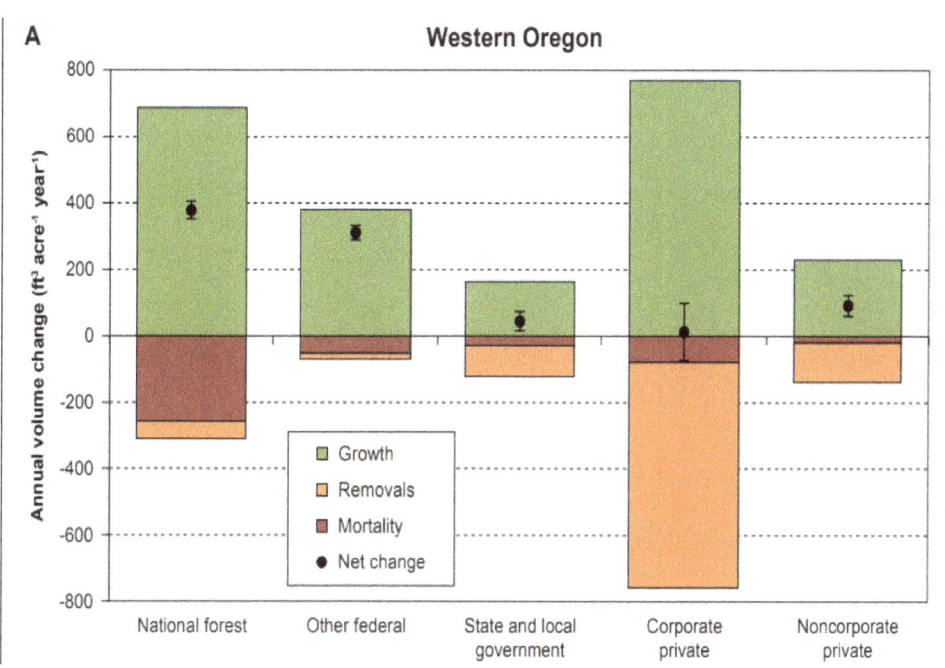

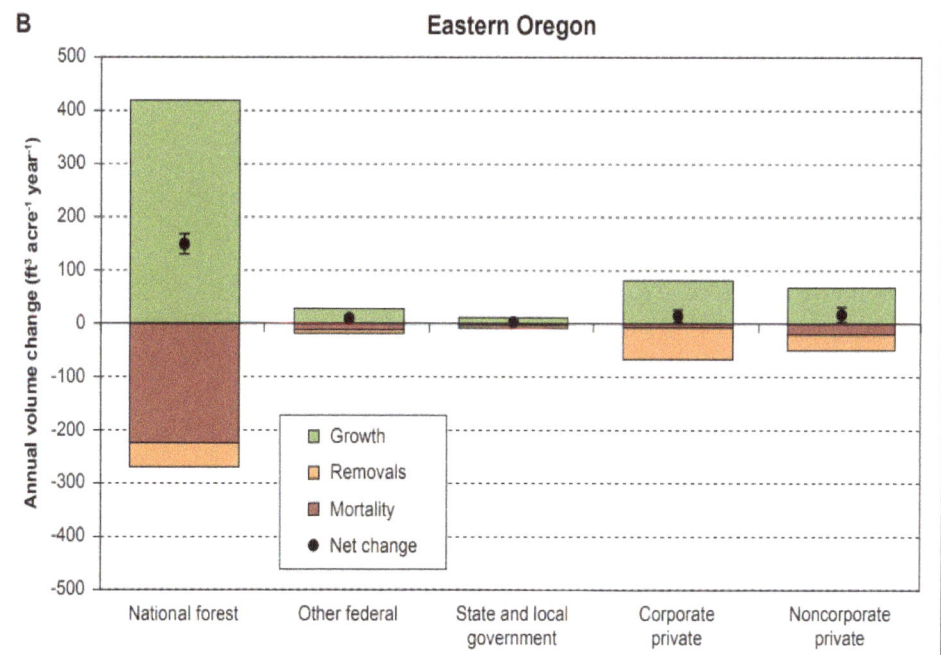

Figure 27—Average annual change in volume (cubic feet per year) of growth, mortality, and removals between 2001–2005 and 2011–2015 by ownership group in (A) western Oregon and (B) eastern Oregon. (John Chase 2018)

For Western and Eastern Oregon, the tables reveal high mortality, high growth rates, and a lack of annual harvesting which is a recipe for future catastrophic-mega-wildfires.

Ron Rommel

An interview with Chris Johnson, VP Operations Forester at Shanda Asset Management Reveals Oregon Forest Facts 2021-2022, (Personal communication, February 11, 2022)

Our greatest resource

"Oregonians won't soon forget Labor Day 2020. Thousands were forced to flee their homes under evacuation orders as massive wildfires that exploded over the holiday weekend raged across western Oregon. Entire towns were destroyed, and more than 4,000 homes were lost. Tragically, nine people were killed by fires that ultimately burned more than 1 million acres. Several factors, including an uncharacteristic wind event, extreme drought conditions and record-low humidity and fuels moisture, came together to make this fire season like no other in recent history…"

Sincerely, Mike Cloughesy, Director of Forestry Oregon Forest Resources Institute Cover photo: inciweb.nwcg.gov

"Nearly half of Oregon is forestland. This forestland has a wide variety of timber productivity levels: high-productivity sites in the Coast Range, which account for 12% of Oregon's forestland; medium-productivity sites in the western Cascades (35% of Oregon forestland); low-productivity sites in eastern Oregon (41% of forestland); and non-productive sites located at high elevations (12% of forestland)."

Ownership:

	Forestland (acres)	Percent of total
U.S. Forest Service	14,093,000	48%
Bureau of Land (BLM)	3,573,000	12%
National Park Service	160,000	1%
Other federal	32,000	<1%
Total federal	17,858,000	60%
State	942,000	3%
County and municipal	187,000	1%
Total state and local	1,129,000	4%
Total government	18,987,000	64%
Large private (>/= 5,000 acres)	6,487,000	22%
Small private (<5,000 acres)	3,702,000	12%
Total private	10,189,000	34%
Native American tribal	480,000	2%
TOTAL FORESTLAND,	29,656,000	100%

Sustainability of Oregon's timber harvest Growth, mortality, and harvest 2005-2015 HARVEST

"On Oregon's private forestland, where most timber harvest happens in the state, the amount of wood harvested each year is about 77 percent of the annual timber growth. About 11 percent of that growth is offset by trees that die from causes such as fire, insects, and disease. <u>On federal lands, only about **8 percent** of the annual timber growth is harvested each year</u>. The amount of timber that dies offsets annual growth by 36 percent. The remainder of the growth, a net change of 56 percent, adds to the volume of standing timber in those forests. High net change in growth isn't always beneficial, however. For example, in federal ponderosa pine and mixed conifer forests in eastern and south-central Oregon, it has created unusually dense forests with stressed trees that are more prone to insect infestation, disease and uncharacteristically severe fire." (Oregon Forest Facts 2021-22 Edition)

(Oregon Forest Resources Institute 2020)

The 2020 and 2021 fire seasons ignited catastrophic wildfires on the larger landscape.

There remains a greater need on the larger landscape scale to include our Federal Forests.

At 8% of the annual timber growth being harvested each year, we have a problem.

Oregon's forest landscape is checker-boarded throughout townships, ranges, and sections. These geographical areas are a mixture of ownership with a majority owned by the federal government. Since the early 1990's, US Forest Service (USFS) managed forestlands have succumbed to a lack of care.

Once again, Congressional Research, the 2006-2015 10-Year Forest Inventory, and Oregon Forest Facts 2021-22 support the need to REDUCE forest fuels on federal timberlands.

High flotation tires (low compaction)

Swing-to Tree Harvester

Tree-to-tree feller buncher

Ground equipment quickly harvests high density forest stands followed by prescribed fire (USDA Forest Service Northern Research Station 2006)

"Typical Central Oregon timber sale is between 4,000-6,000 BF/Acre." (Williams 2022)

We must urgently accelerate fire resilience on federal managed lands.

Our management must be deliberate and consistent and targeted for now and the future. Few virgin giant trees exist that humans squandered and wasted for profit and expansion. Let us begin a new paradigm of forestry, where we use the tools to manage old and new giants. The revenue from the utilization of contracted vendor services can build forest infrastructure.

Past Course of Action Can Aggressively Revitalize the Present

Every year, forests grow and add more and more high-density fuels to the landscape. There is an urgent need for the USFS to vigorously expand a Green Timber Sales Program.

The USFS Green Timber Sales Program is rooted in the past. It can accelerate timber sales while generating revenue. Individual contracts can specify the minimum amount of ground fuels. Increased utilization through site driven silvicultural prescription brings forest products to market while restoring wildfire resilience and preserves habitat.

The revenue from an aggressive Green Timber Sales Program could target:

- Forest restoration with controllable prescribed burns following harvest.
- Maintain strategic access roads.
- Expand surface and ground water sites for fire suppression.
- Provide strategic fire breaks to protect the wildland-urban-interface.
- Hire a full-time staff of well trained and well-paid federal firefighters.

Axe-It-First

As we remain steadfast with NEPA and regulations to protect the forest, we must unite behind fire adaptable plans to save habitats, watersheds, and human infrastructure through the efforts of a consortium of ecologists, biologists, hydrologists, geologists, foresters, and fuels specialists.

As climate change continues to increase the fire risk, we must ACT NOW!

Time Can Be Either Our Friend or Our Enemy

If we fail to act there will be much less time to react to catastrophic-mega-wildfires.

Serving as a Fire Lookout has given me a lot of insight as to what it takes to respond to a wildfire. It takes time to determine the correct azimuth, elevation and to triangulate with adjacent Lookouts to precisely determine the exact location of a fire. "Time is of essence."

Under fire weather atmospheric conditions, you have little time to react and deploy teams.

When a forest is managed for wildfire, you just bought yourself and fire resources TIME.

On January 18, 2022, Agricultural Secretary, Tom Vilsack and USFS Chief, Randy Moore, announced a 10 Year Plan to reduce forest fuel densities on 20 to 30 million acres.

(USDA Secretary, Tom Vilsack and USFS Chief, Randy Moore 2022)

On February 1, 2022, I personally spoke with Oregon US Senator, Jeff Merkley, and discovered that Congress placed a $230 million down payment on the above 10-Year Plan. (Merkley 2022)

Ron Rommel

Fire Patrol Associations Protect Communities from Larger Landscape

Local roots of the Walker Range Fire Patrol Association have a long history in Klamath County, Oregon. According to Walker Range (2021 issue) the owners of Shevlin/Hixon Company, Fremont Land Company, Gilchrist Timber Company, and Ralph E. Gilchrist organized the Walker Range Patrol Association on May 31, 1927. The purpose of the association was to strategically preserve forests from wildfires, insect depredations along with other necessary purposes. (unknown, Walker Range 2021)

Since those early beginnings, Walker Range Fire Patrol Association continues to serve timberland owners and communities under director, R.D. Buell, for the past 47 years. (Hine 2021)

Hine, in issue 2021 said, "nearly 85 percent of wildland fires in the United States are caused by humans, which means the Walker Range took on a daunting three-pronged mission to decrease human-caused fires, aggressively fight and safely manage wildland fires and be an integral member of the community." (Hine 2021)

Buell stated, "Climate change is beating our butt, and we must move with it. Ponds and springs are going dry, so we must haul water more often, and for greater distances." (Hine 2021)

(Author's camera photo of Wise Buys Ads & More and Frontier News)

Walker Range Fire Patrol Association operates the Bald Mountain Fire Lookout

Walker Range Fire Patrol Association serves to protect communities from the larger landscape.

Communities at Risk of Wildfire in the Wildland Urban Interface

(Oregon Department of Forestry 2021)

The above map shows the DARKEST areas rated as HIGHEST RISK to wildfire

Chapter 7
Axe-It-First

Axe-It-First is a plan to balance and adapt today's forests and grasslands to climate change.

It is not a textbook but serves to change the management paradigm and public perception to initiate urgent collective action to prevent annual catastrophic-mega-wildfires. Both historically and presently, the wood products industry is driven by economics. To achieve this new paradigm of forest management on federal lands we must aggressively reduce fuels through logging and biomass utilization. It is a well-known fact, that the structural wood products industry is volatile to the housing market. It is the supply of forest products that drives the timber, building and commodity industries.

During times of crisis or periods of high interest rates the wood products market will either be stymied or halted. Case in point, the 2020 COVID-19 pandemic impacted USA and World economies. In the 1980's interest rates hit 21%. It triggered the timber industry to supply wood chips for pulp and paper. The 2007 Great Recession required flexibility within the industries. If the wood products industry is to remain a key partner for fuels reduction, then there must be an aggressive plan to allow the industry to access federal managed forest resources, in order, to maintain a well-trained industry work force to sustain projects and prevent annual catastrophic fires now and well into the future.

Axe-It-First targets a Plan outlining 5 Essential steps to reduce Catastrophic-Mega-Wildfires.

1. **U.S. Forest Service must expand** their timber sales adjacent to the (WUI) and larger managed landscape.

2. **Reduce ground and ladder fuels** on Federal and adjacent lands near wildland-urban-interface (WUI).

3. **Create strategic fire breaks** to slow fires from non-managed lands to prevent Mega fires in (WUI).

4. **Develop surface and ground** water sources for fire management with an accessible road infrastructure.

5. **Permanently hire USFS** wildland firefighters, train them, pay a living wage and benefits to manage fire.

Ron Rommel

First Essential Step:

US Forest Service must expand timber sales adjacent to the Wildland-Urban-Interface (WUI) and the larger managed landscape.

Wildfire knows **NO** boundaries. All stakeholders must urgently **ACT**.

US Forest Service (USFS) Timber Sale History speaks for itself from 1905-2020

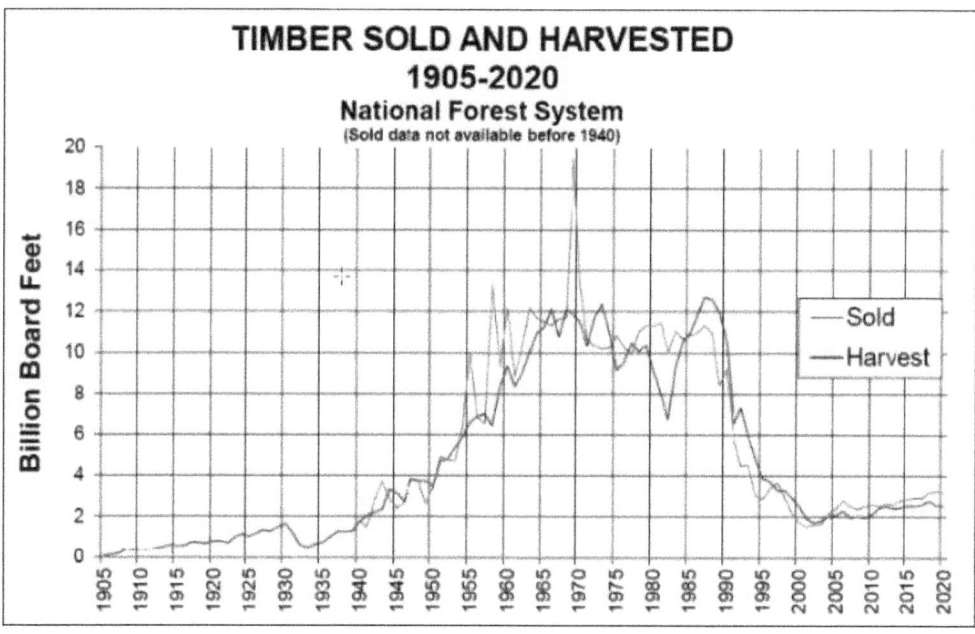

(USFS, FY 1905-2020 National Summary Cut and Sold Data and Graph.pdf, 226KB)
(USDA, Forest Service 2021)

Targeting six billion Board Feet (BF) for each year of utilization would help.

US Forest Service Chief, Tom Tidwell Spoke of Economic Value

"Forests also have economic value, generating wealth through recreation and tourism, through the creation of green jobs, and through the production of wood products and energy. But wood has gotten a bad rap; there is a widespread misconception that building from cement or steel is better for the environment than using wood. We need to dispel those misconceptions, because putting wood to beneficial use is a key strategy for climate mitigation. Wood both stores carbon and replaces more carbon-intensive materials. Lumber is eight times less fossil-fuel-intensive than cement for example —and twenty-one times less fossil-fuel-intensive than steel." (Tom Tidwell 2016)

"Many of the materials we remove to help restore forests have little or no value, but by finding new uses for biomass and small-diameter materials, we can get more restoration work done. For example, researchers at our Forest Products Lab have helped to find ways to use small-diameter materials in cross-laminated timber. The cross-lamination technology creates a stable and structurally sound panel that is used for building components such as floors, walls, ceilings, and more. Completed projects have included the use of these panels for 10-story high-rise buildings!" (Tom Tidwell 2016)

On April 22, 2022, known as; Earth Day, U.S. President Joe Biden signed an Executive Order to Strengthen America's Forests, Boost Wildfire Resilience, and Combat Global Deforestation.

February 2022 down payment increased from $230 million to $8 billion as Bipartisan Infrastructure Law. (whitehouse.gov 2022)

Second Essential Step:

Reduce ground and ladder fuels on Federal and adjacent lands near wildland-urban-interface (WUI)

Oregon's westside and eastside forests contain high density thickets of ladder fuels and volatile ground fuels. Ground fuels, like bitterbrush, manzanita, and ceanothus, contain a high oil content, that quickly ignite and carry a wildfire. For those of us who have worked in forestry, we know that fire shapes the forest ecology by culling out the weakest, and the diseased, while promoting a healthier future forest. Fire is a very necessary component of the forest.

What can you do as a forest manager, as a landowner, as a citizen and as a community?

The challenge that stands before us is addressed on a community basis through **Firewise USA**. Firewise USA has roots dating back to 1896 with the National Fire Protection Association (NFPA). 21st Century Firewise began in 2002 and is based on science... "to help residents get organized, find direction, and take action to increase the ignition resistance of their homes and community Build Defensible Space." (Oregon State Liason (ODF): (Jenna Trentadue) unknown)

Wildland-Urban-Interface (WUI) Benefits from Firewise Communities

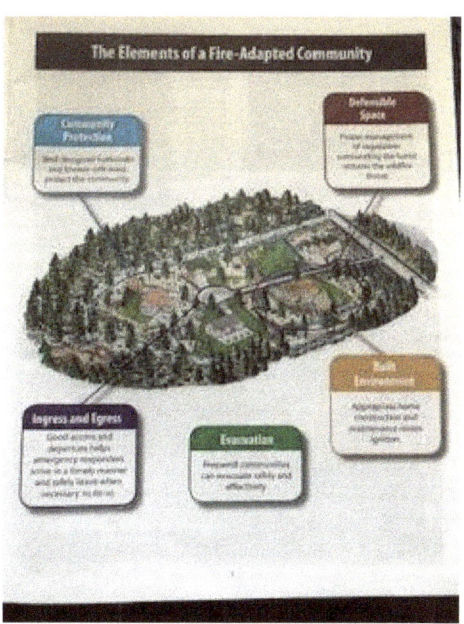

(Oregon State University Extension Service 2015)

Firewise Communities Build Defensible Space Photos Reveal Aggressive Forest Fuels Reduction Near (WUI)

(Authors photos taken in September 2021)

Forest Stewardship Logging is an effective tool

The above photos show the elimination of both ground and ladder fuels through logging and mastication. This proactive model shows how to avoid catastrophic wildfires adjacent to (WUI).

As Federal Forest fuels continue to increase the frequency of catastrophic-mega-wildfires, there will remain a constant threat of wildfire adjacent to the Wildland-Urban-Interface (WUI). To change this paradigm, we must take aggressive action to reduce federal forest fuels.

Collaborative-Partnerships need to trust each other and build solidarity.

We must all deliberately and collectively come together as stakeholders, in partnership, to form a heroic attempt to help forests, grasslands, inclusive habitats, and watersheds adapt to climate change as we attempt to protect human infrastructure and build wildfire resilience.

Failure to reduce the Federal Forest Fuels comes with consequences–

According to Forbes Advisor, issue 2022, "If you live in a geographic area that has an elevated risk of wildfires, you might have a tough time finding affordable homeowners' insurance, in some cases, they decline to write insurance policies for homes in high-risk areas. If you can't find homeowners insurance, you might need to turn to your state's [Fair Access to Insurance Requirements (FAIR) Plan](). But before you can get a policy through a state's FAIR plan, you usually need to be declined by a specific number of insurers first. Insurance through a state FAIR plan is not ideal. Plans can be expensive, have lower coverage and may have restricted coverage, meaning you might have to purchase other coverage types (such as liability insurance) from a private insurance company to get comprehensive homeowners' coverage." (FORBES ADVISOR 2022)

Without homeowner policies, lenders may refuse to finance residential homes.

Oregon's catastrophic-mega-wildfires has IMPACTED the American Insurance Industry,

Farmers and Safeco refuse to write new homeowner policies in Oregon Firewise Communities. ((Steury) 2022)

Firewise focuses on helping communities become fire adapted to withstand wildfires. It takes research and professional forestry leadership to bring about fire-adapted communities.

Dr. Dan Leavell, Ph.D., an Oregon State University Professor with the Klamath Basin Research and Extension Center, believes in partnerships. Dan's work and others like him are providing a path for communities to become proactive at being fire-adapted to survive a catastrophic wildfire.

Why not just fight fire with fire?

Are there any consequences, on the larger landscape, throughout the National Forests?

Primarily, it would require an incredible amount of favorable weather conditions to allow strategic planning to prevent prescribed burns from jumping into adjacent ownerships and exploding. Weather is unpredictable even with the best forecasts. With increasing unpredictability, you can quickly lose control of a large burn. With a finite amount of water and firefighting resources, out-of-control burns can quickly threaten and destroy adjacent forest properties, along with habitats, people's homes, and communities. Harvesting severely fire damaged trees yield little value after a hot burning wildfire. As a log scaler, I culled severely burned wood for failing to meet the scaling rule for pulp/paper chips. Plus, large burns release lots of carbon dioxide (CO_2) that exacerbates climate change.

Fire is a vital component. Employ fire in a manner when you can best control it.

Pete Caligiuri, Forest Ecologist from the Nature Conservancy, indicated anecdotal information suggests that the best outcome for a forest to survive a catastrophic wildfire appears to occur when the forest is managed for resilience by removing ladder fuels, followed by, prescribed fire to reduce ground fuels. On some forest sites prescribed fire may be the only treatment. The Bootleg fire required cross-property boundary collaboration between all stakeholders to control it. (Caligiuri 2022)

Axe-It-First advocates aggressive federal fuel reductions to reduce catastrophic-mega-wildfires. First and foremost, the management of federal lands must step-up.

Utilization must be the first step.

Depending upon the forestland and grasslands sites, the follow up step could be a combination of mechanical and/or prescribed fire to reduce small ladder and ground fuels.

Remember, Congressional Research Service indicated, "Most wildfires are human caused (88% on average from 2016 to 2020), although the wildfires caused by lightning tend to be slightly larger and burn more acreage (55% of the average acreage burned from 2016 to 2020 was ignited by lightning)." (Congressional Research Service 2021)

We must be vigilant!

Third Essential Step:

Create strategic fire breaks to slow fires from non-managed lands to prevent mega fires in wildland-urban-interface (WUI).

- First, utilize through the logging of high-density commercial timber and biomass fuels
- Followed by reducing ground fuels either by mechanical mowing and/or prescribed fire
- Create forest mosaics to breakup continuous fuels from habitats for fire resilience
- Create strategic fire breaks to protect habitats, watersheds, and human infrastructure
- Every site is different. NO one size fits all. No need for perfection, just reduce the Hazard

This is not complicated to plan. Upon ignition, our time is limited.

"Time is of the essence."

Topography, elevation, aspect, selective habitats, and watersheds will dictate the intensity of fuel reductions. We do not need to be perfect, but we must reduce the hazard. Time demands development of an infrastructure that supports a healthy managed forest.

It is critical to have a well-maintained accessible road system for fire suppression equipment. Some road systems may require gates to protect seasonal critical wildlife habitat. Non-essential roads should be decommissioned to protect sensitive habitats and watersheds.

If we embrace logging as a key to promoting healthy forests, we move one step closer toward mitigating high-density forest fuels adjacent to human infrastructure. Allowing a catastrophic wildfire to burn everything to the ground is not an option. Creating forest spatial mosaics and strategic fire breaks goes a long way to prevent catastrophic fires. When we isolate tinder dry forest and grassland fuels, we stand a better chance of slowing down fire and even preventing a catastrophic wildfire.

In my view, the 2020 Lionshead and the P-515 wildfires could have been slowed, perhaps halted, from jumping into Western Oregon if there had been wide strategic fire breaks placed along the eastern flank of the Cascade Crest prior to the east wind event on September 7, 2020.

By aggressively reducing tree density, ladder, and ground fuels, we are closer to fire resilience. "Most people would agree to control fuels." (Smith 2022)

Before you cut, think about the big picture.

Before you grab that chainsaw or jump into that merchandiser, stick to a plan that yields fire resilience when considering habitats, watersheds, and human infrastructure.

The larger managed landscape demands a consortium of specialists to develop a comprehensive plan when implemented provides the needs for all habitats and watersheds while addressing the big picture to prevent catastrophic wildfires. To achieve this, we need to develop a fire management infrastructure.

Fourth Essential Step:

Develop surface and ground water sources to support fire management with an accessible road infrastructure.

In Klamath County, Oregon, water is scarce. We must develop both surface and ground water sources for fire suppression. We need more viable water sources for helicopter dip sites and water tender locations. In addition, we need accessible roads to implement prescribed fire and to fight wildfires. We need a human plan to develop a forest infrastructure.

Fifth Essential Step:

Permanently hire USFS wildland firefighters, train them, pay a living wage and benefits to manage fire.

Increasing forest resilience and adaptability to climate change will require numerous tools and strategies designed to manage the forest. Deploying permanent well trained USFS wildland firefighters to manage fire will improve ecological forest health.

Build Quality Assurance into Forest Management

I propose we take a lesson from my business professor and implement quality assurance into forest management. Dr. Larry Schuetz, Ph. D, taught W. Edward Deming's system known as Plan-Do-Check-Act (PDCA). This system is "…used in business for the control and continuous improvement of processes and products." Originally developed by Walter Andrew Shewhart coined the Shewhart Cycle and later renamed to PDCA. (Wikipedia.org 1900-1993)

Let's transform the forest and grasslands by Planning-Doing-Checking-Acting.

As an observer and as a citizen here's a few things you can do to protect your community:

- Contact Local Fire Protection Associations and Firewise Organizations
- Contact US Senators and Representatives and Demand USFS REDUCE FUELS
- Contact US Forest Service at Office of Communication to express your views
- Remain Proactive through Education and Extension Service Workshops
- Remain the inspiration for others to follow, in order, to achieve fire resilience

As a homeowner living near potential wildfires here are a few wildfire preparedness sites:

https://dfr.oregon.gov/help/Pages/index.aspx Connects an Advocate to YOU with Oregon DFR.

http://www.nfpa.org FIREWISE USA-Wildfire-NFPA

https://firewise.greenoregon.org FIREWISE Oregon, Preparedness

https://wildfire.oregon.gov for Assistance/Common Resources

https://www.oregon.gov/odf/fire/pag for Information & statistics & Active fires

Axe-It-First

map

    https://extension.oregonstate.edu Fire Program-OSU Extension Service

    https://www.klfhp.org/ Klamath Lake Forest Health Partnership Forest Health STARTS WITH YOU.

    http://www.walkerrange.org Walker Range Fire Patrol Association serves North Klamath County

Ron Rommel

Homeowner's Evacuation Check Off List

Your local fire department and/or county sheriff's office typically has a phone app. to keep you updated on the fire in your area. Local Facebook groups also work to keep you informed. Know your fire danger levels at all times.

Level 1 means get ready fire danger is in your area – the below list will help you be prepared.

Level 2 means get set to go – Load level 1 items in your cars (except animals) or move them from your home to a safer area.

Level 3 means GO – you have about 10 minutes to leave the area. Do not go back for items or anything that delays your exit. GO!

Before a fire evacuation it is important to be prepared. Remember, things can be replaced, but lives cannot. Inventorying your possessions, uploading precious family pictures to the cloud or Dropbox, storing important documents, deeds, medical forms when there is no risk of fire makes more sense than attempting to save everything in a moment's notice.

(www.pamalajvincent.com 2022)

Level 1 Fire Evacuation check list:

___Never allow the car gas tanks to fall below ½ a tank.

___ Create a first aid supply kit and keep it in an easily accessible spot

___ Keep a list of important numbers in one place (file, phone, or ?)

___insurance numbers

___credit card and bank numbers (accounts and contact numbers)

___hotels (if you have pets, know ahead of time which ones allow them)

___medical records

___ATM cards, photo id, birth certificates, passports, marriage licenses, social security cards, pet id tags, any paperwork you may need in the event everything burns.

___Important papers box – insurance forms, titles, life insurance, wills, etc.

___ family valuables, photo albums, jewelry, irreplaceable items. You never go back for these items once you have evacuated, so have them ready prior to leaving.

___Know your evacuation route—there may be long lines

___Pack two weeks' worth of medications (you may be gone longer)

___the CDC recommends cleaning supplies, masks, etc. in the event of disease and staying in a shelter. Public shelters do not often allow pets unless they are service animals.

___Cash – set aside enough money to eat and pay for a hotel for two weeks. Be sure to include small bills in case stores are out of electricity and exact cash is vital.

___Create a family plan to stay in contact and a central meeting place.

___Pack personal items and clothing for each person – consider the weather and comfort.

___Phone charger in your car and one in your bag. However, if power is out, you may have a tougher time making contact – (value of a central meeting place with family)

___ Check on your neighbors

___ Large animals need to be removed early- know where you can take them.

___ Pack pet supplies, food, water, meds, bowls, collars, leashes, bedding, and carriers.

___ Pack pet vaccine records (many places will require proof of vaccinations)

___ Take photos of your possessions on a phone for possible insurance claims

___ Move propane tanks, combustibles away from your home.

___ Water grass and foliage if you have time.

___ If you have security cameras, charge the batteries. For more ideas

Axe-It-First

Conclusion

Our history of deforestation, slaughtered wildlife, and decimated Indigenous peoples have led to this moment in time. When we learn from our mistakes, we grow stronger and wiser to become an inspiration for future generations. We learn good choices yield positive results.

For us to alter the future outcome of the forest, its inhabitants, and values, we must look at ourselves as the elephant in the room. Earth's human population exploded in 1800.

Distant centuries ago, earth's human population was estimated to have been around three hundred million. Much further back in time, lesser clans of humans left a small impact as they roamed the Earth.

Today, 7.9 billion humans impact Earth's climate and its resources. Barry Commoner, biologist, ecologist, Ph.D. professor once said, "The Earth has not only experienced a population explosion, but also and more meaningfully, a civilization explosion." (Commoner 1974)

The expansion of America has taken its toll on the ecosystem. Our cities, towns, and communities continue to expand development for more and more people. Far too many people are eager to expand and enhance economic growth at the expense of ignoring the long-term view of how we impact both the current and future ecosystems.

We must grasp the notion that Earth is finite with finite resources.

It will take us, within our societies to face our political, cultural, and religious beliefs. Every child born and every man and woman who lives, impacts the Earth. We, as a species, have accelerated climate change and its impact upon the Earth.

To lower the human impact, it may take cleaner technology and lower human birth rates.

Most of all, it will require each one of us to realize, we have the capacity to effect measurable change and examine our place on this Earth, to save Earth from ourselves.

People care about our natural world, so let yourself be that person to rescue our resources! Let's be the catalyst to effect change and be the inspiration to implement effective solutions!

Ron Rommel

Axe-It-First

Glossary

Basal Area per Acre: Average amount of square feet per acre occupied by tree stems. (https://www.mdwfp.com/media/4194/basal_area_guide.pdf) It is used to for calculating the Scribner Board Feet per acre. (Author provided application)

CLORPT: To form soil, it takes Climate, Organisms, Relief, Parent material over time.

Cubic foot (CF): (1) cubic foot = 12" x 12" x 12" = 1728 cubic inches. (Author provided definition)

DOI US Department of Interior.

Forbs Any herbaceous flowering plant, like herbs.

Hectares (ha) International measurement. It is 2.471 acres per US acres.

Ladder fuel Ladder fuel is a firefighting term used to describe live or dead vegetation allowing a fire to climb up a tree canopy or hillside. Typical ladder fuels include tree branches, shrubs, and tall grasses both living and dead. They fuel the fires from the ground up to the largest trees in a forest. A fuel break is horizontal and vertical spacing between vegetation (live or dead) or other flammable materials reducing risk of fire's ability to spread.

NEPA National Environmental Protection Act. Became law in 1970.

Pleistocene: Pleistocene is known as the Great Ice Age. It began 2,580,000 to 11,700 years ago. (https://en.wikipedia.org/wiki/Pleistocene)

Scribner Board Feet (BF): Scribner Board Feet (BF) is derived from Scribner Log Rule. A diagram rule where (1) BF = 12" x 12" x 1" = 144 square inches. Scribner Board Feet (BF) is the standard of measurement when determining the measured gross and net volume of logs from timber harvest. The economic value is measured by Scribner (BF). Logs are bought and sold using Scribner (BF). (Author provided definition)

Bibliography

Andres Schmidt, Daniel Leavell, John Punches, Marco A. Rocha Ibarra, James S. Kagan, Megan Creutzburg, Myrica McCune, Janine Salwasser, Cara Walter, Carrie Berger. 2022. "A quantitative wildfire risk assessment using a modular approach of geostatistical clustering and regionally distinct valuations of assets-A case study in Oregon." Edited by Lalit Kumas Sharma Zoological Survey of India. PLOS ONE (Open Access) (Oregon State University) 32. Accessed March 10, 2022. https://journals.plos.org/plosone/article?id=10.1371/journal.pone.0264826.

Dr. Dan Leavell Ph.D., Forestry Professor at OSU authorized the sharing of this research document with Author, Ron Rommel.

Baley, Randy, interview by Ron Rommel. 2022. Oregon Department of Forestry, forester (February 18).

Barnas, Ron, and US Forest Service Law Enforcement Officer Clackamas Ranger District of Mt. Hood National Forest. 1980. "Olallie Lake Resort." Smokey Bear Plaque for Fire Prevention. Oregon.

Caligiuri, Pete, interview by Ron Rommel. 2022. Forest Ecologist for Nature Conservancy fo Bend, Oregon (January 24).

Chris Johnson, interview by Ron Rommel. 2022. V.P. Operations Forester, Shanda Asset Management, LLC. (February 11). an original member of the Deschutes Collaborative Forest Project

Commoner, Barry. 1974. The Closing Circle, Nature, Man & Technology. Bantam. Vol. 9. 9-12 vols. New York City, New York: Bantam Books, Inc. Alfred A. Knopf, Inc. Accessed March 2022.

Congressional Research Service. 2021. Wildfire Statistics. October 4. Accessed March 21, 2022. https://sgp.fas.org/crs/misc/IF10244.pdf.

David C. Powell, Silviculturist. 2013. Silvicultural Activities: Description and Terminology. Pacific Northwest Region, Umitilla National Forest USDA Forest Service. February. Accessed February 2022. https://www.fs.usda.gov/Internet/FSE_DOCUMENTS/stelprdb5413732.pdf.

David J. Mildrexler, Logan T. Berner, Beverly E. Law, Richard A. Birdsey and William R. Moomaw. 2020. frontiers in Forests and Global Change, Forest Management. November 5. Accessed February 2022. https://www.frontiersin.org/articles/10.3389/ffgc.2020.594274/full#:~:text=10.3389%2Fffgc.2020.594274-,Large%20Trees%20Dominate%20Carbon%20Storage%20in%20Forests%20East%20of%20the,the%20United%20States%20Pacific%20Northwest&text=Large%2Ddiameter%20trees%2.

Deschutes Collaborative Forest Project. unknown. Deschutes Collaborative Members. unknown unknown. Accessed April 2, 2022. deschutescollaborativeforest.org/deschutes-collaborative-members-2/.

Diaz, Cindy McCain and Nancy. 2002. Field Guide to the Forested Plant Associations of the Westside Central Cascades of Northwest Oregon. Pages 178-181.Washington, DC 20250: United States Department of Agriculture Forest Service Pacific Northwest Region Technical Paper R6-NR-ECOL-

Axe-It-First

TP-02-02.

Dr. Dan Leavell, interview by Ron Rommel. 2022. Ph.D. Forestry Professor at Oregon State University (March 9).

Encyclopedia.org, Colorado. 2002. 2002 Missionary Ridge Fire. June 9. Accessed March 2022. https://coloradoencyclopedia.org/article/missionary-ridge-fire.

EPA, and Environmental Protection Agency. n.d. Air Quality Index. Accessed 2022. https://www.epa.gov/wildfire-smoke-course/wildfire-smoke-and-your-patients-health-air-quality-index.

FORBES ADVISOR. 2022. What to Know About Wildfire Insurance. Edited by Jason Metz. February 28. Accessed April 12, 2022. https://www.forbes.com/advisor/homeowners-insurance/wildfires/.

Franklin, J.F. and R.H. Waring 1979. Evergreen Coniferous Forests of the Pacific Northwest-Oregon State. Science.org; volume 204, page 1381 in original research paper written as PDF. June 29. Accessed December 2021. http://andrewsforest.oregonstate.edu.

unknown. Geology, Soils & Climate. EOLA-AMITY HILLS Winegrowers. Accessed April 16, 2022. https://eolaamityhills.com/geology-soils-climate/Geology, Soils & Climate.

Hine, Andrea. 2021. Walker Range Committed to Reducing Wildfires, October 5-11: 18. Accessed October 5-11, 2021. walkerrange.org.

John Chase, Jeremy Fried, Sarah Jovan, Katherine Mercer, Andrew Gray, David M. Bell, Sara Loreno, Todd Morgan. 2018. Oregon Forest Resources, 2006-2015: 10-Year Forest Inventory and Analysis Report.Gen.Tech.Rep. PNW-GTR-971. Edited by Marin Palmer, Olaf Kuegler and Glenn, tech.eds.2018 Christensen. Forest Service U.S. Department of Agriculture. October. Accessed February 2022. https://www.fs.fed.us/pnw/pubs/pnw_gtr971.pdf.

Kathy Sarns, Alaska, Artist's rendition of the Fire Triangle – Source: USFWS. 2006. Artist's rendition of the Fire Triangle – Source: USFWS Alaska. USFWS Alaska granted author Ron Rommel access 2022. https://www. redzone.co/2016/02/17/wildfire-101-the-fire-triangle-and-the-fire-tetrahedron/.

Katu.com. 2021. Katu.com Photo of Bootleg Fire. Accessed March 2022. https://www.google.com/search?rlz=1C1RUCY_enUS787US788&source=univ&tbm=isch&q=Katu.com+file+photo+of+2021+Bootleg+Fire&fir=wCnLCXc6I-Hq-M%252C9fqvLH5yJkcD4M%252C_%253BT-PdF1S4IkJRDM%252C9fqvLH5yJkcD4M%252C_%253BX_GDvFu-Q4rFeM%252CtobGrAwxzeaw_M%252C_%253B.

Klamath Lake Forest Health Partnership (KLFHP). 2020. Forest Health STARTS WITH YOU. Accessed March 9, 2022. https://www.klfhp.org/.

Library.org Photo of Missionary Ridge Aircraft Deploying Fire Retardant

New York Times. 2021. New York Times photo of unmanaged and managed forests 2021 Bootleg fire. Access March 2022.

OregonLive.com. 2020. Echo Mountain Fire near Otis, Oregon. Accessed March 2022. https://www.google.com/search?q=Oregon+live.com+of+Otis%2C+Oregon+2020+Wildfire&tbm=isch&ved=2ahUKEwjYs5bprYr3AhVuFTQIHd0cDxcQ2-cCegQIABAA&oq=Oregon+live.com+of+Otis%2C+Oregon+2020+Wildfire&gs_lcp=CgNpbWcQAzoECAAQQzoHCAAQsQMQQzoLCAAQgAQQsQMQgwE6BQgAEIAEOg.

Oregon Forest Resources Institute, Oregon Forest Facts 2021-22 Edition, 2020.

Robbins, William G. unknown. Timber Industry Photos. unknown unknown. Accessed February 2022. https://www.oregonencyclopedia.org/articles/timber_industry/#.YlI6KcjMKyI.

2020. Scientific Consensus: Earth's Climate Is Warming. Accessed April 4, 2022. https://climate.nasa.gov/scientific-consensus/.

Smith, Nick, interview by Ron Rommel. 2022. Founder of Healthy Forests Communities in 2013 (February 17).

(Steury), Sandra Michelle Oliver, interview by Ron Rommel. 2022. Sandra Steury Insurance Agency, Inc. (February 15).

Stowe, David, interview by Ron Rommel. 2022. Chair of Bend, Oregon Chapter of the Sierra Club and participate of the Deschutes Collaborative Forest Project and the Central Oregon Shared Stewardship Alliance (February 8).

Text by Aileen Agnew, 2021. 17th Century. The Devil and the Wilderness. Accessed November 2021. www.mainememory.net.

the-Journal.com. 2002. the Journal.com/articles/firefighter-battle-slow-moving-wildfire-on-Missionary-ridge/. Accessed March 2022. https://www.google.com/search?rlz=1C1RUCY_enUS787US788&source=univ&tbm=isch&q=the-Journal.

Tom Tidwell, Chief of U.S. Forest Service 2016. 2016. Tom Tidwell's Speech to the World Conservation Congress, Honolulu, HI. September 4. Accessed February 2022. https://www.fs.usda.gov/speeches/state-forests-and-forestry-united-states-1.

unknown. unknown. Radiant Heat Transfer. Accessed October 2021.

USDA Department of Agriculture, Forest Service. unknown. Partnership Resource Center. Accessed February 2022. https://www.fs.usda.gov/main/prc/legal-administrativeresources/collaboration-law.

USDA Forest Service Northern Research Station. 2006. North Central Region WEB-BASED FOREST MANAGEMENT GUIDELINES (Photos). May 26. Accessed April 11, 2022. https://www.nrs.fs.fed.us/fmg/nfmg/fm101/silv/p3_harvest.html.

USDA Secretary, Tom Vilsack and USFS Chief, Randy Moore. 2022. Secretary Vilsack Announces New 10 Year Strategy To Confront Wildfire Crisis. January 18. Accessed January 18, 2022. https://www.fs.usda.gov/news/releases/secretary-vilsack-announces-new-10-year-strategy-confront-wildfire-crisis.

USDA, Forest Service. 2021. USFS, FY 1905-2020 National Summary Cut and Sold Data and Graphs. June 29. Accessed February 2022. https://www.fs.fed.us/forestmanagement/documents/sold-harvest/documents/1905-2020_Natl_Summary_Graph.pdf.

Vincent, Pamala J. Emergency Preparedness list @ www.Pamalajvincent.com

2021. Walker Range F.P.A. History. Webmaster Services by: Code3Creative. Accessed October 5-11, 2021. walkerrange.org.

whitehouse.gov. 2022. FACT SHEET: President Biden Signs Executive Order to Strengthen America's Forests, Boost Wildfire Resilience, and Combat Global Deforestation. April 22. Accessed April 22, 2022. https://www.whitehouse.gov/briefing-room/statements-release/ 2022/04/22/fact-sheet-president-biden-signs-executive-order-to-strengthen-americas-forests-boost-wildfire-resilience-and-combat-global-deforestation/.

Wikipedia, the free encyclopedia. 2021. 2021 Bootleg Wildfire near Beatty, Oregon. July-August. https://en.wikipedia.org/wiki/Bootleg_Fire.

Wikipedia.org. 2020. 2020 Oregon Wildfires. (https://en.wikipedia.org/wiki/2020_Oregon_wildfires).

Wikipedia, the free encyclopedia. 2003. Cedar Fire. October 25. https://en.wikipedia.org/wiki/Cedar_Fire).

Wikipedia, fuel ladder, ladder fuels, accessed may 2022

Wikipedia.org. 2021. Dominance (ecology). June. Accessed February 2022. https://en.wikipedia.org/wiki/Dominance_(ecology).

—. 1900-1993. W. Edward Deming (Overview). Accessed April 2022. https://en.wikipedia.org/wiki/W._Edwards_Deming.

William W. Bergoffen, Forest Service Retired. 1976. 100 Years Of Federal Forestry. Vol. Agriculture Information Bulletin No. 402. Stock No. 001-000-03668-8 vols. Washington, D.C. 20402, Maryland and Virginia: U.S. Government Printing Office. Accessed January 2022.

Williams, John, interview by Ron Rommel. January 2022 President of Quicksilver Contracting Company (January 17).

Ron Rommel speaks on targeting and preventing catastrophic mega wildfires.

If you would like Ron to speak to your organization, contact him at
rommelron192@gmail.com

www.ingramcontent.com/pod-product-compliance
Lightning Source LLC
Chambersburg PA
CBHW062103290426
44110CB00022B/2699